PERENNIAL MATH

Tournament Tests 2

Grades 5-6

PERENNIAL MATH TOURNAMENT TESTS

These are a collection of tests from past Perennial Math Tournaments. Each tournament has two rounds of testing. Each student takes an Individual Test during the first round and then students combine with their teammates to take a Team Test during the second round. The solutions included in this book are only a short sample of the many possible solutions that could be used to solve each problem.

INDIVIDUAL TESTS GRADES 5-6

30 minute practice tests for individual students

Grade 5-6 Individual Event (30 Minutes)

1) Mike has $50 made up of dimes and Bill has $50 made up of quarters. How many more coins does Mike have than Bill?

2) Fill in the boxes to complete this subtraction problem What is the sum of these four missing numbers?

	□	3	1	□
−	3	□	□	3
	3	3	3	3

3) Square ABCD has a perimeter of 16 cm. What is the area of the circle that is inside? Use 3.14 for π

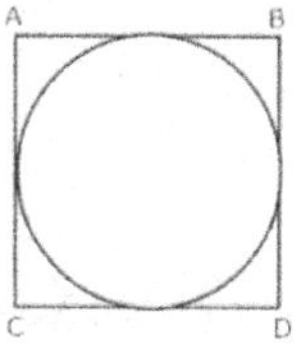

4) Bob's team scores an average of 41 points per game in its first four games and an average of 40 points per game in its first five games. How many points did his team score in its fifth game?

5) If

A
B D
C

means (A + D) x (B + C), Express

16
19 17
18

in simplest form.

6) If 2 pears weigh as much as 3 peaches, and 2 peaches weigh as much as 30 grapes, then how many pears weigh as much as 90 grapes?

7) Mitch has some rabbits and pigeons in his barn. He counted 30 heads and 76 legs. How many rabbits did he have?

8) Three pencils and a pen costs 78 cents. Two pencils and two pens cost 84 cents. How much does 1 pen cost?

9) Every year Jimmy has put pennies in the jar on his birthday. He has put in the same number of pennies as his age each year. He now has $1.05 in the jar. How old is he?

10) It takes 10 men to build 150 barrels a day. How many men does it take to build 60 barrels in one day?

PERENNIAL MATH

11) $9 + 3 \times 2^2 - 1 =$

12) 2352 hours = ____ weeks

13) $1.095 \times 23.52 =$
Round to the nearest tenth.

14) What is the ones digit in this equation:
$9 \times 8 \times 7 \times 6 \times 5 \times 4 \times 3 \times 2 =$

15) If the sum of two different numbers is 60, their product at most can be _______?

INDIVIDUAL ANSWER DOCUMENT (Grades 5-6)

SCORE

Team # __________ School ______________________________

Student Name ______________________________

Student Number D - __________

1. __________
2. __________
3. __________
4. __________
5. __________
6. __________
7. __________
8. __________
9. __________
10. __________
11. __________
12. __________
13. __________
14. __________
15. __________

PERENNIAL MATH

Grade 5-6 Individual Event Solutions

1) $50 of dimes is 500 dimes. $50 of quarters is 200 quarters. 500 – 200 = 300 more coins.

2)The sum of the four missing numbers is 7 + 9 + 8 + 6 =30.

$$\begin{array}{r} \boxed{7}\ \boxed{3}\ \boxed{1}\ \boxed{6} \\ -\ \boxed{3}\ \boxed{9}\ \boxed{8}\ \boxed{3} \\ \hline \boxed{3}\ \boxed{3}\ \boxed{3}\ \boxed{3} \end{array}$$

3) Since the perimeter of the square is 16 cm, then the length of each side is 4cm. The diameter of the circle would also be 4cm and radius of 2cm. The area of a circle is π times radius squared. 3.14 x 2^2 = 3.14 x 4 = 12.56 cm^2.

4) Bob's team scores an average of 41 points per game in its first four games and an average of 40 points per game in its first five games. To determine how many points his team scored in its fifth game first find the total points for both sets of games and then subtract. If the average of the first four games was 41 then the total points was 41 x 4 = 164. The average of the first five games is 40 so the total points is 5 x 40 = 200. Now subtract the total points to find the score of the fifth game: 200 - 164 = 36.

5) If

$$\begin{array}{ccc} & A & \\ B & & D \\ & C & \end{array}$$

means (A + D) x (B + C), Express

$$\begin{array}{ccc} & 16 & \\ 19 & & 17 \\ & 18 & \end{array}$$

in simplest form.

(16 + 17) x (19 + 18) = 33 x 37 = 1221

6) If 2 pears weigh as much as 3 peaches, and 2 peaches weigh as much as 30 grapes, then how many pears weigh as much as 90 grapes? 90 grapes is going to be equal to 6 peaches. If 2 pears weigh as much as 3 peaches, then double the pears and 4 pears weigh as much as 6 peaches. So 4 pears weigh as much as 90 grapes.

7) Mitch has some rabbits and pigeons in his barn. He counted 30 heads and 76 legs. Make a chart to see how many rabbits he has. Start with half of each animal and notice that you need to decrease the number of rabbits. Mitch has 8 rabbits.

Rabbits	15	10	9	8
Pigeons	15	20	21	22
Total heads	30	30	30	30
Rabbit feet	60	40	36	32
Pigeon feet	30	40	42	44
Total feet	90	80	78	76

PERENNIAL MATH

Grade 5-6 Individual Event Solutions

8) Three pencils and a pen costs 78 cents. Two pencils and two pens cost 84 cents. To solve for the cost of 1 pen use algebra. Let c represent pencils and p represent pens.
If 2c + 2p = 84, then divide both sides by 2 and c + p = 42. Now subtract this from the first equation and solve.

$$\begin{array}{r} 3c + 1p = 78 \\ \underline{-\ c + \ \ p = 42} \\ 2c \qquad = 36 \end{array} \text{ or } c = 18$$

If c equals 18 and c + p = 42, then 18 + p = 42 or p = 42 – 18 = 24. The cost of p (1 pen) is 24 cents or $0.24.

9) Every year Jimmy has put pennies in the jar on his birthday. He has put in the same number of pennies as his age each year. He now has $1.05 in the jar. So do the addition for each year until you get 1.05: .01 + .02 + .03 + .04 + .05 + .06 + .07 + .08 + .09 + .10 + .11 + .12 + .13 + .14 = 1.05. Jimmy is 14 years old.

10) One man can make 15 barrels. 60/15= 4; Therefore, it takes 4 men 1 day to make 60 barrels.

11) $9 + 3 \times 2^2 - 1 =$
Use the order of operations to solve.
9 + (3 x 4) - 1 = 21 - 1 = 20

12) Since there are 24 hours in a day and 7 days in a week, you can find the number of weeks by dividing 2352 hours by 24. That will give you 98 days. 98/7 = 14 weeks

13) 1.095 x 23.52 = 25.7544. This is 25.8 when rounded to the nearest tenth.

14) The equation 9 x 8 x 7 x 6 x 5 x 4 3 x 2 = 362,880 The ones digit is 0. A quick way to look at it would be to notice that 5 times any even number will always yield a 0 in the ones digit.

15) If the sum of two numbers is 60, their product at most can be _______?
The greatest product can only be achieved by using the two largest numbers possible and multiplying them. The two largest different numbers are 31 and 29. 31 x 29 = 899.

INDIVIDUAL ANSWER KEY

(Grades 5-6)

Instructions for Scoring:

- ***Mark CORRECT answers by circling "O" the question number, and INCORRECT answers with an "X" . Each correct answer is worth one point. Count the circles (correct answers) and write that number in the "SCORE" box.***

CORRECT → (1.) ______

INCORRECT → X. ______

1. **300**
2. **30**
3. **12.56**
4. **36**
5. **1221**
6. **4**
7. **8**
8. **24 or .24**
9. **14**
10. **4**
11. **20**
12. **14**
13. **25.8**
14. **0**
15. **899**

Grade 5-6 Individual Event (30 Minutes)

1) What is the sum of the first 9 odd prime numbers?

2) What number goes in the square to make the equation correct?

$$18 + \square = 26 - \square$$

3) This figure contains 6 congruent squares. How many inches are in the perimeter of the figure if its area is 96 inches^2?

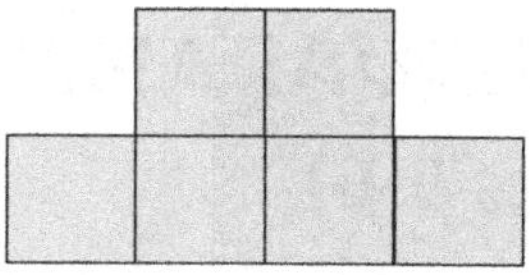

4) It takes Mr. Maples 20 minutes to cut a long plank into 5 short pieces. At the same rate how long would it take Mr. Maples to cut another identical plank into 9 pieces?

5) Millie is reading a book. The sum of the two open page numbers is 385. What is the next page number in the book?

6) Drew organizes his pile of coins by putting them in bags. If he puts coins into 3 bags, he has 2 coins left over. If he puts coins into 4 bags, he has 2 left. If he puts coins into 5 bags, he has 2 left over. What is the least number of coins that Drew can have altogether?

7) One banana and one pear weigh the same as 14 strawberries. Two pears weigh the same as 10 strawberries. Using this scale, how many strawberries weigh the same as one banana?

8) Andrew, Hal and Kevin went to the orchard to pick apples. Andrew picked one third of what Hal picked. Hal picked 4 times as many as Kevin did. If Kevin picked 27 apples, how many apples did Andrew pick?

9) Farmer Brown has some chickens and cows. There are 30 heads and 76 legs in all. How many chickens does Farmer Brown have?

10) There are 44 students in a class. Twenty-seven students like chocolate ice cream and twenty-nine like vanilla ice cream. Fifteen like both. How many students like neither chocolate nor vanilla?

PERENNIAL MATH

11) $12 - 2 \times 3 + 12 =$

12) $61{,}875 \div 125 =$

13) $1\frac{4}{6} \div 2\frac{2}{7} =$

Answer must be reduced fraction

14) What is the average of these numbers: 33, 35, 37, 39, 41, 43, 45

15) $11 + 22 + 33 + 44 + 55 + 56 + 67 + 78 + 89 =$

INDIVIDUAL ANSWER DOCUMENT
(Grades 5-6)

SCORE

Team # __________ School ______________________________

Student Name ______________________________________

Student Number D - ______________

1. ________	6. ________	11. ________
2. ________	7. ________	12. ________
3. ________	8. ________	13. ________
4. ________	9. ________	14. ________
5. ________	10. ________	15. ________

PERENNIAL MATH

Grade 5-6 Individual Event Solutions

1) 3 + 5 + 7 + 11 + 13 + 17 + 19 + 23 + 29 = 127 .

2) 18 + [4] = 26 - [4]

3) The figure contains 6 congruent squares with total area of 96 inches^2, so the area of one square is 96inches^2/6 = 16 inches^2. The length of the sides of the squares are 4 inches each. There are 12 sides going around the perimeter so the perimeter is equal to 12 x 4 inches = 48 in.

4) In order to cut the plank into 5 pieces, Mr. Maples needs to make 4 cuts. Each cut takes 5 minutes. To cut a plank into 9 pieces, Mr. Maples needs to make 8 cuts. Therefore it takes 40 minutes to make 8 cuts.

5) 192 and 193 = 385. The next page is 194.

6) The number of coins Drew has is a "multiple" of 3, 4 and 5 plus 2. The least common multiple of 3, 4 and 5 is 3 x 4 x 5 = 60; Therefore the minimum number of coins Drew has is 62.

7) Since 2 pears weigh the same as 10 strawberries, 1 pear weighs the same as 5 strawberries. Therefore, 1 banana's weight is the same as 14 – 5 = 9 strawberries.

8) Hal picked = 27 x 4 = 108 apples;
Andrew picked = 108 ÷ 3 = 36 apples.

9) 22 chickens and 8 cows.
Make a table and you will notice that the number of legs decrease by 2 when you add a chicken and take away a cow.

Chickens	15	16	17	22
Cows	15	14	13	8
Heads (30)	30	30	30	30
Legs (76)	90	88	86	76

10) Use a Venn Diagram.
The number of students who like chocolate = 27
Chocolate only = 27 - 15 = 12
The number of students who like vanilla = 29
Vanilla only = 29 - 15 = 14
The number of students who like neither = 3

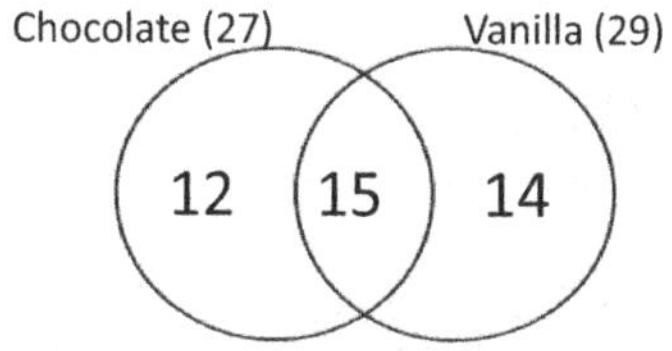

Neither
44- (12 + 15 + 14) = **3**

11) 12 - 2 x 3 + 12 = 12 – (6) +12 = 18

12) 61,875 ÷ 125 = 495

13) $1\frac{4}{6} \div 2\frac{2}{7}$ = 10/6 x 7/16 = 70/96 = 35/48

14) The average of these numbers: 33, 35, 37, 39, 41, 43, 45 is 39. It is the middle number.

15) Use smart addition: (11 + 89)+(22 + 78)+(33 + 67)+(44 + 56)+55 =455

PERENNIAL MATH

INDIVIDUAL ANSWER KEY

(Grades 5-6)

Instructions for Scoring:

- ***Mark CORRECT answers by circling "O" the question number, and INCORRECT answers with an "X". Each correct answer is worth one point. Count the circles (correct answers) and write the number in the "SCORE" box.***

CORRECT → (1.) ________

INCORRECT → X. ________

1. 127
2. 4
3. 48
4. 40
5. 194
6. 62
7. 9
8. 36
9. 22
10. 3
11. 18
12. 495
13. 35/48
14. 39
15. 455

Grade 5-6 Individual Event (30 Minutes)

1) A runner burns 1200 calories per hour. In 75 minutes how many calories will the runner burn?

2) For what value of X does
$1 + 2 + 3 + 4 + \cdots + X = 120$?

3) In this cryptarithm each letter represents a different digit. What is the sum of E + F?

$$\begin{array}{r} A34 \\ B1C \\ +\ 53D \\ \hline E4F3 \end{array}$$

4) If the digits 6, 5, 4, 3, 2, 1 are written in that order repeatedly, what will be the 2000th digit?

5) A sandwich 15 inches long is cut into 6 pieces of equal length. What is the sum of the lengths of four of these pieces?

6) A barrel is one-fifth full of pecans. We add 25 pecans and the barrel it is now 1/4 full. How many more pecans must be added to make the barrel 1/2 full?

7) In the fraction $\frac{p}{q}$: The numerator p may equal 1, 2, 3, or 6. The denominator q may equal 1, 2, or 4. How many different values are possible for the number $\frac{p}{q}$?

8) Tim is building a three-feet wide sidewalk around a rectangular garden that is 60 feet by 70 feet. How many square feet will be in the area of the sidewalk?

9) How many different ways can 35 cents be made from any combination of has pennies, nickels, dimes, and quarters?

10) There are 29 girls in the soccer club. To raise money for their tournament 12 girls washed cars, 13 girls participated in a bake sale, and 6 girls raked leaves. Three girls were involved in all three activities. Five girls did both the car wash and bake sale. Four girls did both the bake sale and raking. Three girls did both raking and the car wash. How many girls did none of the fundraising activities?

11) 9 - 3 x 8 + 12 =

12) ____ - (-1) = -2

13) $\frac{98.3}{10.1} = \frac{a}{95}$
(Round to nearest tenth)

14) 1 + 2 + 3...+ 25 =

15) Find the LCM of 12, 15, and 18.

INDIVIDUAL ANSWER DOCUMENT
(Grades 5-6)

SCORE

Team # __________ School ______________________________

Student Name______________________________________

Student Number_ D - ____________

1. ____________
2. ____________
3. ____________
4. ____________
5. ____________
6. ____________
7. ____________
8. ____________
9. ____________
10. ____________
11. ____________
12. ____________
13. ____________
14. ____________
15. ____________

PERENNIAL MATH

Grade 5-6 Individual Event Solutions

1) A runner burns 1200 calories per hour equals 1200/60 or 20 calories per minute. In 75 minutes the runner burn will burn 75 x 20 = 1500 calories.

2) 1+ 2 + 3 + 4 + 5 + 6 + 7 + 8 + 9 + 10 + 11 + 12 + 13 + 14 + 15 = 120.   X = 15.

3) The letters C and D must add up to 9 and the sum of the ones column carries so F = 8. Now moving on to the hundreds column because the tens column does not carry A + B + 5 = E4 or 14 because E4 cannot be 24. So E = 1 and F = 8 and the sum of E + F = 9.

$$\begin{array}{r} A34 \\ B1C \\ +\ 53D \\ \hline E4F3 \end{array}$$

4) If the digits 6, 5, 4, 3, 2, 1 are written in this order repeatedly, the 2000th digit will be 5, because there are 6 numbers in this series and 2000 ÷ 6 = 333 with a remainder of 2. The second digit in the series is 5.

5) A sandwich 15 inches long is cut into 6 pieces of equal length. 15 ÷ 6 = 2.5. Each piece is 2.5 inches. The length of 4 of these pieces is 2.5in x 4 = 10 inches.

6) A barrel is one-fifth full of pecans. We add 25 pecans to the barrel and it is now 1/4 full. To find out how many more pecans must be added to make the barrel 1/2 full, solve for how many pecans make up 1/4 a barrel. Use algebra and solve for x which stands for the number of pecans in the barrel. Then find out how many pecans are in 1/4 of a barrel.

1/5x + 25 = 1/4x

1/5 x – 1/5x + 25 = 1/4x – 1/5 x   (subtract 1/5x from both sides)

25 = 1/20 x   (reduce)

500 = x   (multiple both sides by 20)

So, 1/4 x = 500/4 = 125, therefore 125 more pecans would need to be added to make the barrel half full.

7) In the fraction $\frac{p}{q}$: The numerator p may equal 1, 2, 3, or 6. The denominator q may equal 1, 2, or 4. How many different values are possible for the number $\frac{p}{q}$?

1/1, 1/2, 1/4, 2/1, 2/2, 2/4, 3/1, 3/2, 3/4, 6/1, 6/2, 6/4. When these fractions are simplified: 1, 1/2, 1/4, 2, 1, 1/2, 3, 1 ½, 3/4, 6, 3, 1 ½ . Eliminate any duplicate values and 1, 1/2, 1/4, 3/4, 2, 3, 6/4 and 6 remain. There are 8 different possible values for fraction $\frac{p}{q}$.

Grade 5-6 Individual Event Solutions

8) Tim is building a three-feet wide sidewalk around a rectangular garden that is 60 feet by 70 feet. To calculate square feet of the area of the sidewalk, find the area of the garden with the sidewalk and then subtract the area of the garden. The area of the garden and sidewalk is (60 + (2 x 3)) x (70 + (2 x 3)) = 66 x 76 = 5016ft^2. The garden is 60 x 70 = 4200ft^2. So, the sidewalk area is 5016ft^2 - 4200ft^2 = 816ft^2

9) Make a chart to find the possible coin combinations for .35. There are 24 ways.

Quarters	Dimes	Nickels	Pennies
1	1	0	0
1	0	2	0
1	0	1	5
1	0	0	10
0	3	1	0
0	3	0	5
0	2	3	0
0	2	2	5
0	2	1	10
0	2	0	15
0	1	5	0
0	1	4	5
0	1	3	10
0	1	2	15
0	1	1	20
0	1	0	25
0	0	7	0
0	0	6	5
0	0	5	10
0	0	4	15
0	0	3	20
0	0	2	25
0	0	1	30
0	0	0	35

10) Use a Venn Diagram. All regions total the 29 girls in the soccer club.

Region C = 12 – (3 + 2 + 0) = 7
Region B = 13 – (3 + 2 +1) = 7
Region R = 6 – (0 +3 +1) = 2

To solve for who did none of the activities subtract all the regions from 29.
29 – (7 + 2 + 3 + 1 + 7 + 0 + 2) = 7.
Seven girls didn't participate in fundraising.

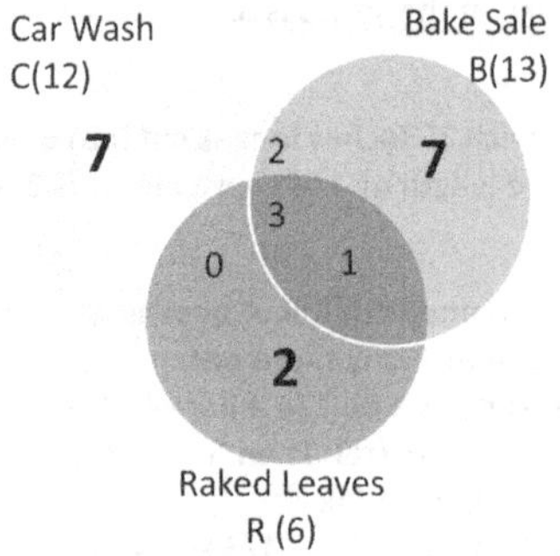

11) 9 - 3 x 8 + 12 =
Use the order of operations PEMDAS to solve.
9 - (3 x 8) + 12 = 9 - 24 + 12 = -3

12) ____ - (-1) = -2
-3 + 1 = -2

13) $\frac{98.3}{10.1} = \frac{a}{95}$

Use cross multiplication and solve
98.3 x 95 = 10.1a
9338.5 = 10.1a
9338.5÷10.1 = a therefore, a = 924.60396039. Round this to 924.6

14) Use smart addition and group pairs together.
(1 + 24) +(2 + 23) + (3 + 22) + (4 +21) + (5 + 20) + (6 + 19) + (7 + 18) + (8 + 17) + (9 + 16) + (10 + 15) + (11 + 14) + (12 + 13) + 25 = 13 x 25 = 325

15) Find the LCM of 12, 15, and 18.
Factorize the numbers:
12 = 2 x 2 x 3
15 = 3 x 5
18 = 2 x 3 x 3
Count the number of times each prime factor appears and multiple.
LCM = 2^2 x 3^2 x 5 = 4 x 9 x 5 = 180

PERENNIAL MATH

INDIVIDUAL ANSWER KEY

(Grades 5-6)

<u>*Instructions for Scoring:*</u>

- ***Mark CORRECT answers by circling "O" the question number, and INCORRECT answers with an "X". Each <u>correct answer</u> is worth one point. Count the circles (<u>correct answers</u>) and write that number in the "SCORE" box.***

CORRECT → (1.) ________

INCORRECT → X. ________

1. **1500**	6. **125**	11. **-3**
2. **15**	7. **8**	12. **-3**
3. **9**	8. **816**	13. **924.6**
4. **5**	9. **24**	14. **325**
5. **10**	10. **7**	15. **180**

Grade 5-6 Individual Event (30 Minutes)

1) What is the sum of the first 10 prime numbers?

2) According to this diagram it takes 7 toothpicks to make a row with 3 triangles. How many toothpicks does it take to make a row with 67 triangles?

3) If 3 darts land on this dart board, how many different point totals are possible?

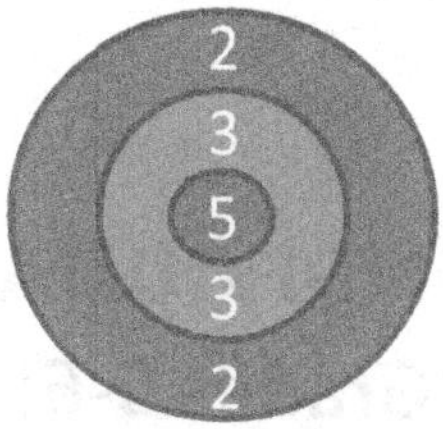

4) A toy shop makes tricycles and four-wheel wagons. The owner counted 30 total vehicles and 102 wheels. How many wagons were in the toy shop?

5) The sum of three consecutive numbers is 81. What is the first of the three numbers?

6) On a 25 question math test, Mrs. Smith gives four points for every correct answer and takes away one point for each incorrect answer. If Bob scores 65 on the test, how many did he get correct?

7) Mack had a big bag of gumdrops. He gave 1/8 of them to his mother and then gave 1/2 of what was left to his sister. He then gave 2/7 of what was left to his dad and had 25 gumdrops left. How many gumdrops did he have to start with?

8) How many two-digit even numbers can be made using the digits 0, 1, 2, 3, 5, 7, 8 if each digit can appear only once in each number?

9) On Monday the temperature was -10 degrees. On Tuesday it rose nine degrees. On Wednesday it dropped 15 degrees and on Thursday it rose seven degrees. What was the temperature on Thursday?

10) James has dimes, nickels, and pennies in his piggy bank. How many different ways can he make 57 cents?

11) $8 - 4 \times 2 + 12 =$

12) $12{,}210 \div 555$

13) $3\frac{5}{6} \div \frac{3}{4} =$

14) What is the average of these numbers: 23, 25, 27, 29, 31, 33, 35

15) $23 + 35 + 47 + 53 + 65 + 77 =$

INDIVIDUAL ANSWER DOCUMENT
(Grades 5-6)

SCORE

Team # __________ School ______________________________

Student Name ____

Student Number _________________

1. ____________ 6. ____________ 11. ____________

2. ____________ 7. ____________ 12. ____________

3. ____________ 8. ____________ 13. ____________

4. ____________ 9. ____________ 14. ____________

5. ____________ 10. ____________ 15. ____________

PERENNIAL MATH

Grade 5-6 Individual Event Solutions

1) The sum of the first ten prime numbers is 2 + 3 + 5 + 7 + 11 + 13 + 17 + 19 + 23 +29 = 129.

2) There is a pattern that exists. Let N represent the number of triangles. The formula for the number of toothpicks is 2N + 1. Therefore, for 67 triangles you need 2(67) + 1 = 135 toothpicks.

3) Make an organized list. There are 7 different point totals that can be made from 3 darts landing.

Dart 1	Dart 2	Dart 3	Point Total
5	5	5	15
5	5	4	14
5	5	3	13
5	4	4	13
5	4	3	12
5	3	3	11
4	4	4	12
4	4	3	11
4	3	3	10
3	3	3	9

4) Make a table and you will notice that the number of legs decrease by 1 when you add a wagon and take away a tricycle. There are 12 wagons and 18 tricycles for a total of 102 wheels.

Tricycles	15	16	17	18
Wagons	15	14	13	12
Total Wheels	105	104	103	102

5) 81 ÷ 3 = 27 which is the average or middle number. That would make 26 the first number in the set { 26, 27, 28}.

6) Use a table to solve the problem.

Problems Correct	Points	Points Taken Away	Score
25	100	0	100
24	96	1	95
23	92	2	90
22	88	3	85
21	84	4	80
20	80	5	75
19	76	6	70
18	72	7	65

7) Work backwards. Mack ended with 25 which was 5/7. So dad received 10 gumdrops which was 2/7. So he had 35 gumdrops after he gave his sister half. So if half is 35, he had 70 after he gave his mother 1/8 of them which would make him having 7/8. If 70 is 7/8, then he gave mom 10. 70 + 10 is 80 gumdrops to start with. Work it forwards to check: 80 x 1/8 = 10 70 gumdrops left. 70 x 1/2 = 35 gumdrops left. 35 x 2/7 = 25 gumdrops left.

PERENNIAL MATH

Grade 5-6 Individual Event Solutions

8) There are 16 two-digit even numbers that can be made using the digits 0, 1, 2, 3, 5, 7, 8 where each digit can appear only once in each number. 10, 12, 18, 20, 28, 30, 32, 38, 50, 52, 58, 70, 72, 78, 80, 82

9) Students can draw out a thermometer or use integers to solve: Monday (-10) + 9 (Tuesday) = -1 degrees. -1 -15 (Wed) = -16 degrees. -16 + 7 (Thurs) = -9 degrees.

10) There are 42 combinations he can use to make 57 cents.

Dimes	Nickels	Pennies
5	1	2
5	0	7
4	3	2
4	2	7
4	1	12
4	0	17
3	5	2
3	4	7
3	3	12
3	2	17
3	1	22
3	0	27
2	7	2
2	6	7
2	5	12
2	4	17
2	3	22
2	2	27
2	1	32
2	0	37
1	9	2
1	8	7
1	7	12
1	6	17
1	5	22
1	4	27
1	3	32
1	2	37
1	1	42
1	0	47
0	11	2
0	10	7
0	9	12
0	8	17
0	7	22
0	6	27
0	5	32
0	4	37
0	3	42
0	2	47
0	1	52
0	0	57

11) 8 - 4 x 2 + 12 =
Use the order of operations PEMDAS to solve.
8 - (8) + (12) = 12

12) 12,210 ÷ 555 = 22

13) $3\frac{5}{6} \div \frac{3}{4}$ = 23/6 x 4/3 = 46/9 = $5\frac{1}{9}$

14) The average of these numbers: 23, 25, 27, 29, 31, 33, 35 is 29 because it is the middle number.

15) 23 + 35 + 47 + 53 + 65 + 77 =
Use smart addition and group pairs together.
23 + 77 + 35 + 65 + 47 + 53 =
100 + 100 + 100 = 300

INDIVIDUAL ANSWER KEY
(Grades 5-6)

Instructions for Scoring:

- *Mark CORRECT answers by circling "O" the question number, and INCORRECT answers with an "X". Each correct answer is worth one point. Count the circles (correct answers) and write that number in the "SCORE" box.*

CORRECT → (1.) ________

INCORRECT → X. ________

1. **129**	6. **18**	11. **12**
2. **135**	7. **80**	12. **22**
3. **7**	8. **16**	13. $5\frac{1}{9}$ or $5.\overline{1}$
4. **12**	9. **-9**	14. **29**
5. **26**	10. **42**	15. **300**

TEAM TESTS GRADES 5-6

20 minute practice tests for teams
Each member gets a copy of the test, but the team only receives one answer document

Grade 5-6 Team Event (20 Minutes)

1) The sum of five consecutive even numbers is 150. What is the fifth number?

2) In this 3 x 3 square there is one dark square in the center. How many white squares would border a square that had nine dark squares in the center?

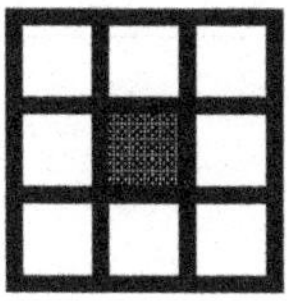

3) Each number in Column 2 is derived from adding 2 numbers from Column 1 and subtracting the third number. Each of the numbers in Column 3 has been derived in the same way from the numbers in Column 2. Solve for the numbers in Column 4. How much is Y?

Column 1	Column 2	Column 3	Column 4
32	40	24	X
35	34	36	Y
37	30	44	Z

4) The math club is doing a fundraiser buying and selling lollipops. They buy 6 lollipops for a dollar and they sell each one for 40 cents. They made a $56 profit by the end of the day. How many dozen lollipops did they sell?

5) If 5 moons equal 4 suns and 12 suns equal 5 stars, then 30 moons equal how many stars?

6) Angie, Betty, and Cindy went shopping. Angie and Betty had $76 together. Betty and Cindy had $81 together. Angie and Cindy had $79 together. How much money did Cindy have?

7) Grandma can make a blanket in 3 days, but it takes Aunt Jenny 4 days to make the same blanket. If they work together, how many days will it take them to make 35 blankets?

8) Annie and Marla each had 32 beads. Annie gave Marla half of her beads. Then Marla gave Annie ¼ of all the beads she had now. After that Marla had how many more beads than Annie?

9) Shelly has 8 coins which are made up of dimes, nickels and pennies. If she must have at least 1 of each type of coin, how many different possible sums could she have?

10) Bob earned $78 raking leaves. He earned twice as much as Tom and Eddie combined. If Eddie earned $17, how much did Tom earn?

TEAM ANSWER DOCUMENT

(Grades 5-6)

SCORE

Team # __________ School ______________________________

Student Names______________________________________

____

Write answers clearly. Each correct answer is worth one point.

1. __________

2. __________

3. __________

4. __________

5. __________

6. __________

7. __________

8. __________

9. __________

10. __________

PERENNIAL MATH

Grade 5-6 Team Event Solutions

1) If the sum of 5 consecutive even numbers is 150, then the average of those 5 numbers is 150/5 = 30. That makes 30 the middle number and the five numbers are 26, 28, 30, 32, and 34. Therefore 34 is the 5th number.

2) A square that had nine dark squares in the center would have a 3 x 3 square in the center and be surrounded by 16 white squares on the border.

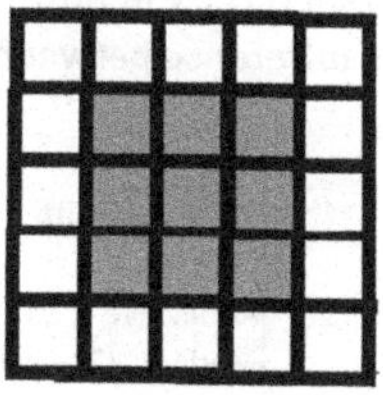

3) The first number in each column is derived from subtracting the first number from the sum of second and third number of the previous column. The second number is derived from subtracting the second number from the sum of the first and third number of the previous column. The third number in each column is derived from subtracting the third number from the sum of the first and second number of the previous column. Therefore Y equals (24 + 44) – 36 = 32

4) For each dozen lollipops they paid $2 and they sold them for (12 x .40 = $4.80). So they made 4.80 – 2.00 = 2.80 profit per dozen. Ten dozen = 10 x 2.80 = 28.00 profit. 56.00 is twice that amount or 20 dozen.

5) If 5 moons equal 4 suns and 12 suns equal 5 stars, then 30 moons equal how many stars. First convert the moons to suns. The ratio of moons to suns is 5:4; therefore 30 moons is equal to 6 x 4 = 24 suns. Since 12 suns equal 5 stars, we have 2 times that so 24 suns equal 10 stars. 30 moons is equal to 10 stars.

6) Use Algebra with A=Angie, B=Betty, and C=Cindy

$$\begin{aligned} A + B \quad\quad &= 76 \\ B + C &= 81 \\ \underline{A + \quad\quad C} &\underline{= 79} \\ 2A + 2B + 2C &= 236 \end{aligned}$$

Divide by 2 and A + B + C = 118. Now you can substitute the original equation (A + B) into this and deduce that 76 + C = 118. So, C = 42; Cindy has $42

Grade 5-6 Team Event Solutions

7) In 12 days they can make 7 blankets working together. (Grandma can make 4 and Aunt Jenny can make 3). To make 35 blankets (7 x 5) will take 12 x 5 or 60 days.

8) Create a chart to keep track of how many beads each person has after each transaction. Subtract to find the difference between Marla and Annie. She has 8 more beads.

	Annie	Marla	
They start with 32 each.	32	32	
Annie give 1/2 to Marla, 32/2 = 16 beads	16	48	32-16 = 16 32 + 16 = 48
Marla gives 1/4 to Annie 48/4 = 12	28	36	16 + 12 = 28 48 -12 = 36
Subtract to find how much more Marla had than Annie			36 -28 = 8

9) Create an organized list and start with the most amount of dimes you can have and work your way down. There are 21 different possible sums she can have with the 8 coins.

Dimes	Nickels	Pennies	Sum
6	1	1	0.66
5	2	1	0.61
5	1	2	0.57
4	3	1	0.56
4	2	2	0.52
4	1	3	0.48
3	4	1	0.51
3	3	2	0.47
3	2	3	0.43
3	1	4	0.39
2	5	1	0.46
2	4	2	0.42
2	3	3	0.38
2	2	4	0.34
2	1	5	0.30
1	6	1	0.41
1	5	2	0.37
1	4	3	0.33
1	3	4	0.29
1	2	5	0.25
1	1	6	0.21

10) Bob earned $78 and he earned twice as much as Tom and Eddie combined, so they earned 78/2 = $39. If Eddie earned $17, then Tom earned $39-17 = $22.

PERENNIAL MATH

TEAM ANSWER KEY

(Grades 5-6)

<u>Instructions for Scoring:</u>

- ***Mark CORRECT answers by circling "O" the question number, and INCORRECT answers with an "X" . Each correct answer is worth one point. Count the circles (correct answers) and write that number in the "SCORE" box.***

CORRECT ⟶ (1.) ________

INCORRECT ⟶ X. ________

1. **34**
2. **16**
3. **32**
4. **20**
5. **10**
6. **42**
7. **60**
8. **8**
9. **21**
10. **22**

Grade 5-6 Team Event (20 Minutes)

1) What is the sum of the first seven prime numbers greater than 31?

2) In the sequence below. How many toothpicks will it take to make to make a line of 41 connecting triangles?

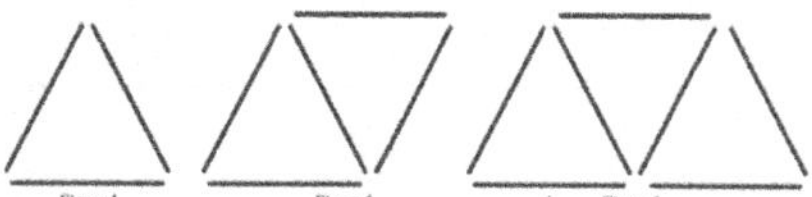

Figure 1 Figure 2 Figure 3

3) How many rectangles of any size are in this diagram that do not include gray?

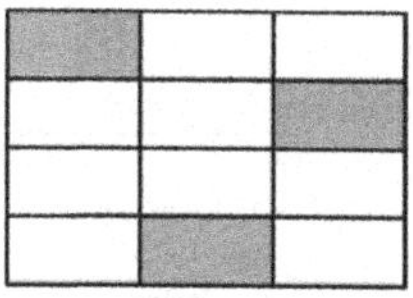

4) Max has a bag of 50 marbles. 20% of the marbles are blue. He gave Eric half of the marbles that were not blue. How many total marbles did Max have left?

5) In the theater there are 15 rows of seats. Each row has the same number of seats. If we were to take away 3 rows, then the theater could hold the same number of people but the remaining 12 rows would need 4 extra seats. How many seats are in the theater?

6) Mark was elected class president. He received 3 votes for every 2 votes that Neil got. No one else ran. However, if 6 of the people who voted for Mark had voted for Neil instead, Mark would have received 2 votes for every 3 that Neil would have gotten. How many people voted?

7) Arnold, Barry and Chris go to the arcade. When they pull out their money they discover that Arnold and Barry have 27 dollars together. Barry and Chris have 31 dollars combined. And Arnold and Chris have 34 dollars together. How much money does Chris have by himself?

8) In this magic square, numbers can be placed in the boxes so that the sum of the three numbers in each row, each column, and in each diagonal is the same amount. What value should the middle box have?

	2	
8	?	4
3		5

9) Sharon went to a store and spent one-third of her money. She went to a second store where she spent one-third of what remained, and then had $16 when she left. How much money did she have to begin with at the first store?

10) Jill is four years older than her friend Pam. Pam is three years older than Amy. In 4 years, Jill will be twice as old as Amy. How old is Pam now?

PERENNIAL MATH

TEAM ANSWER DOCUMENT
(Grades 5-6)

SCORE

Team # __________ School ______________________________

Student Names______________________________

____

Write answers clearly. Each correct answer is worth one point.

1. __________

2. __________

3. __________

4. __________

5. __________

6. __________

7. __________

8. __________

9. __________

10. __________

PERENNIAL MATH

Grade 5-6 Team Event Solutions

1) The sum of first seven prime numbers greater than 31 is 37 + 41 + 43 + 47 + 53 + 59 + 61 = 341

2) You could draw these out and count or if you make a chart you can see the relationship of Triangles to Toothpicks and deduce the formula that Toothpicks = 2 x Number of Triangles +1. Therefore, for 41 triangles you need 2 x 41 + 1 = 83 toothpicks.

Number of Triangles	1	2	3	41
Number of Toothpicks	3	5	7	83

3) How many rectangles of any size are in this diagram that don't include gray? Make a chart. Count shapes going across, then down.

Type	How Many
Made of 1 shape	9
Made of 2 shapes	9
Made of 3 shapes	3
Made of 4 shapes	1
Total	22

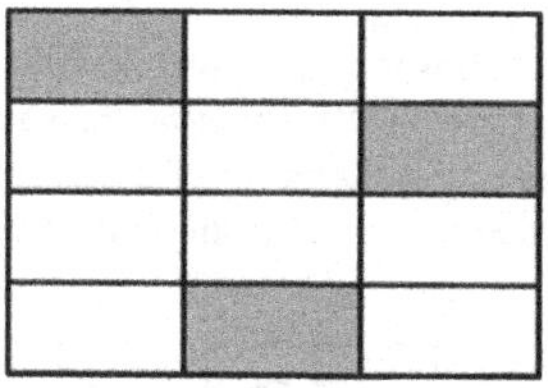

4) Max had a total of 50 marbles. If 20% are blue, then 80% are not blue. 80% of 50 is 40 marbles. He gave half of the 40 marbles to Eric. That would be 20. So now Max has 10 + 20 = 30 total marbles left.

5) You could draw this out and have 15 rows. Since you are moving 3 rows out to leave 12 rows and those 12 rows are each receiving 4 chairs added to them, then there were 48 chairs moved from those 3 rows. You can deduce that each row had to have had 16 chairs in it to start with since you moved 48 chairs and they made up 3 rows. 15 original rows times 16 seats equal 240 seats in the theater.

6) Set up a table where the first scenario has a 3:2 ratio and the second scenario subtracts 6 votes from Mark and adds 6 votes to Neil. Look for numbers in the second scenario that have a 2:3 ratio. In the original election, Mark received 18 votes to Neil's 12 votes so there were 30 students who voted.

Mark's Votes	3	6	9	12	15	18	21	24
Neil's Votes	2	4	6	8	10	12	14	16
Mark's Votes - 6	-3	0	3	6	9	12	15	18
Neil's Votes + 6	8	10	12	14	16	18	20	22

Grade 5-6 Team Event Solutions

7) Use Algebra

$$\begin{array}{lcl} A + B & & = 27 \\ & B + C & = 31 \\ A + & \quad C & = 34 \\ \hline 2A + 2B + 2C & & = 92 \end{array}$$

Divide by 2 and A + B + C = 46. Now you can substitute the original equations and deduce that 27 + C = 46, So C = 19. Substitute C in and B + 19 = 31, therefore B is 12. Substitute 12 in for B and get A + 12 = 27, therefore A is 15. Chris has the most $19.

8) The middle box is shared second row and second column so that you can do algebra. So row (8 + ? + 4) = (2 + ? + X). Simplify and you get (12) = (2 + X). X = 10 and now you can solve for what the sums should be for all sides and diagonals. 3 + 10 + 5 = 18. Since all rows and columns are equal you can solve for the middle box. 18 = 2 + ? + 10. The middle box is 6.

	2	
8	?	4
3	X	5

	2	
8	?	4
3	10	5

	2	
8	6	4
3	10	5

9) Sharon spent 1/3 of her money in store 1 and had 2/3 left. In the second store she spent 1/3 of 2/3 of her original amount. In both stores she spent 1/3+(1/3 x 2/3) 3/9 + 2/9 = 5/9 of her money. Clearly she had 4/9 of her money left which was equal to $16. Then 1/9 of her money was $4. She originally had 9 x 4 or $36.

	Left 2/3	Spent 1/3	Amount When Entering Store
Second Store	16	8	24
First Store	24	12	36

10) Make a chart to compare ages now and in 4 years. Pam is 6 years old now.

Now			In 4 Years			
Amy	Pam (Amy + 3)	Jill (Pam + 4)	Amy + 4	Jill + 4	Pam + 4	Check Is Jill = 2 * Amy
1	4	8	5	8	12	~~10~~
2	5	9	6	9	13	~~12~~
3	6	10	7	10	14	14

TEAM ANSWER KEY

(Grades 5-6)

Instructions for Scoring:

- ***Mark CORRECT answers by circling "O" the question number, and INCORRECT answers with an "X" . Each correct answer is worth one point. Count the circles (correct answers) and write that number in the "SCORE" box.***

CORRECT → (1.) ________

INCORRECT → X. ________

1. 341
2. 83
3. 22
4. 30
5. 240
6. 30
7. 19
8. 6
9. 36
10. 6

Grade 5-6 Team Event (20 Minutes)

1) Brian is reading a book. The sum of the two open page numbers is 355. What is the next page number in the book?

2) This rectangular pool measures 12 by 20 feet. There is a 2ft sidewalk that goes around the pool. What is the area of the entire sidewalk and pool?

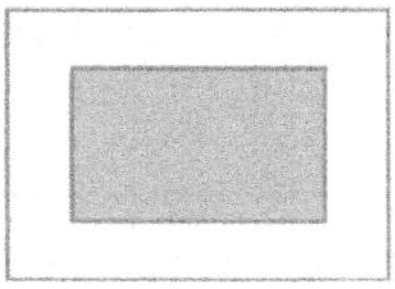

3) Find the surface area of this rectangular prism.

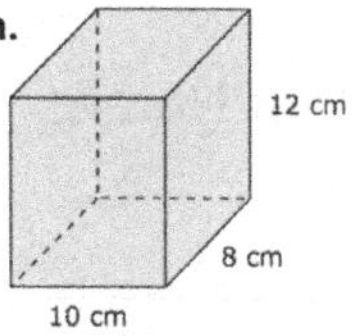

4) Andrew, Howie and Ken went to pick apples. Andrew picked one third of what Howie picked. Howie picked 4 times as many as Ken did. If Ken picked 27 apples, how many apples did Andrew pick?

5) When Ron put his pennies in stacks of 3, he has 2 pennies left over. If he puts his pennies into stacks of 4, he has 2 pennies left over. If he puts his pennies into stacks of 5, he has 2 pennies left over. What is the least number of pennies that Ron can have?

6) There are 44 students in a class. Twenty-seven students like cookies and twenty-nine like vanilla ice cream. Sixteen like both. How many students like neither cookies nor vanilla ice cream?

7) Andrea and Bella have 27 candy bars together. Bella and Cameron have 31 candy bars combined. And Andrea and Cameron have 34 candy bars together. How many candy bars does Cameron have?

8) Marilyn was elected captain of the cheer squad. She received 3 votes for every 2 votes that Shelia got. However, if 6 of the people who voted for Marilyn had voted for Shelia instead, Marilyn would have received 2 votes for every 3 that Shelia would have received. How many people voted?

9) In the theater there are 16 rows of seats. Each row has the same number of seats. If we were to take away 4 rows, then the theater could hold the same number of people but the remaining 12 rows would need 3 extra seats. How many seats are in the theater?

10) It takes Mike 20 minutes to cut a log into 5 pieces. At the same rate how long would it take him to cut another log into 9 pieces?

PERENNIAL MATH

TEAM ANSWER DOCUMENT
(Grades 5-6)

SCORE

Team # __________ School ______________________________

Student Names ______________________________________

____

Write answers clearly. Each correct answer is worth one point.

1. __________

2. __________

3. __________

4. __________

5. __________

6. __________

7. __________

8. __________

9. __________

10. __________

PERENNIAL MATH

Grade 5-6 Team Event Solutions

1) To find the page numbers, first find the average of 355/2 = 177.5. So the page numbers are 177 + 178= 355. The next page would be 179

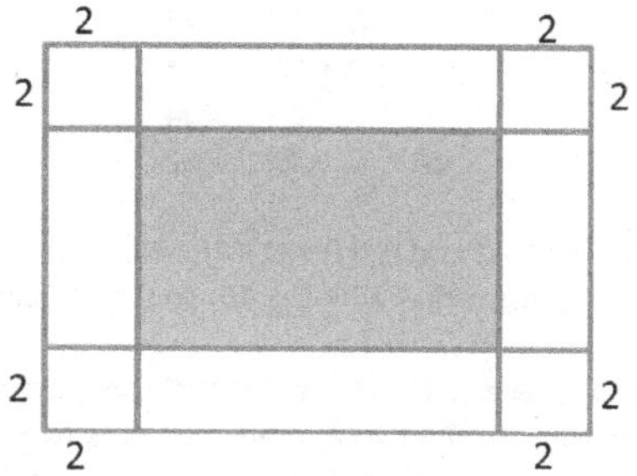

2) The area of the entire shape can be calculated by adding 2 to the ends of the length and the width.
(12 +2 + 2) x (20 + 2 + 2) = (16) x (24) = 384m^2

3) To find the surface area of a rectangular prism it will be equal to the twice the area of each of the three sides. The area of 1st side (top) is 10cm x 8cm= 80cm^2. The second side (sides) would be 8cm x 12cm= 96cm^2. The area of the 3rd side (front) would be 10cm x 12cm= 120cm^2. 2(80cm2 + 96cm^2 + 120cm^2) = 592cm^2

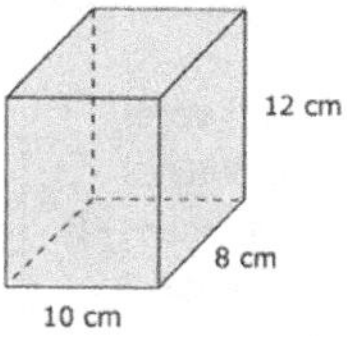

4) Work backwards.

Howie picked four times as many as Ken or 27 x 4 = 108 apples;
Andrew picked 1/3 of Howie or 108 ÷ 3 = 36 apples.

5) The number of pennies Ron has is a "multiple" of 3, 4 and 5 plus 2 (the two pennies left over). The least common multiple of 3, 4 and 5 is 3 x 4 x 5 = 60; Therefore the minimum number of pennies Ron has is 62.

6) Use a Venn Diagram.
The number of students who like cookies = 27
Cookies only = 27 - 16 = 11
The number of students who like vanilla ice cream = 29
Vanilla ice cream only = 29 - 16 = 13

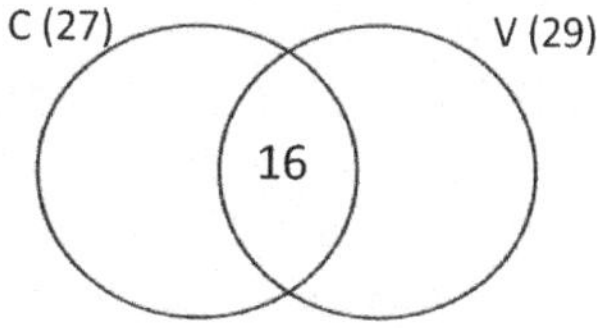

Neither 44 - (11 + 16 + 13) = **4**

Grade 5-6 Team Event Solutions

7) Use Algebra with A=Andrea, B=Bella, and C=Cameron

$$
\begin{array}{rl}
A + B \qquad & = 27 \\
B + C & = 31 \\
\underline{A + \qquad C} & \underline{= 34} \\
2A + 2B + 2C & = 92
\end{array}
$$

Divide by 2 and $A + B + C = 46$. Now you can substitute the original equation $(A + B)$ into this and deduce that $27 + C = 46$, So $C = 19$; Cameron has 19 candy bars

8) Set up a table where the first scenario has a 3:2 ratio and the second scenario subtracts 6 votes from Marilyn and adds 6 votes to Shelia. Look for numbers in the second scenario that have a 2:3 ratio. In the original election, Marilyn received 18 votes to Shelia's 12 votes so there were 30 students who voted.

Marilyn's Votes	3	6	9	12	15	18	21	24
Shelia's Votes	2	4	6	8	10	12	14	16
Marilyn's Votes - 6	-3	0	3	6	9	12	15	18
Shelia's Votes + 6	8	10	12	14	16	18	20	22

9) You could draw this out and have 16 rows. Since you are moving 4 rows out to leave 12 rows and those 12 rows are each receiving 3 chairs added to them, then there were 36 chairs moved from those 4 rows. You can deduce that each row had to have had 9 in it to start with since you moved 36 chairs and they made up 4 rows. 16 original rows times 9 seats equal 144 seats in the theater.

These 4 rows will move and add 3 chairs to the remaining 12 rows, so $3 \times 12 = 36$.

Since there were 4 rows moved $36/4 = 9$ seats in each row at the start.

10) In order to cut the log into 5 pieces, Mike needs to make 4 cuts. Each cut takes $20 / 4 = 5$ minutes. To cut a log into 9 pieces, Mike needs to make 8 cuts. Therefore it takes (8cuts x 5min) or 40 minutes to make 8 cuts.

TEAM ANSWER KEY

(Grades 5-6)

Instructions for Scoring:

- ***Mark CORRECT answers by circling "O" the question number, and INCORRECT answers with an "X" . Each correct answer is worth one point. Count the circles (correct answers) and write that number in the "SCORE" box.***

CORRECT → (1.) ________

INCORRECT → ~~1.~~ ________

1. **179**
2. **384**
3. **592**
4. **36**
5. **62**
6. **4**
7. **19**
8. **30**
9. **144**
10. **40**

Grade 5-6 Team Event (20 Minutes)

1) What is the sum of the first 7 prime numbers?

2) Study the first two triangles. Using this pattern what is the missing number on triangle 3?

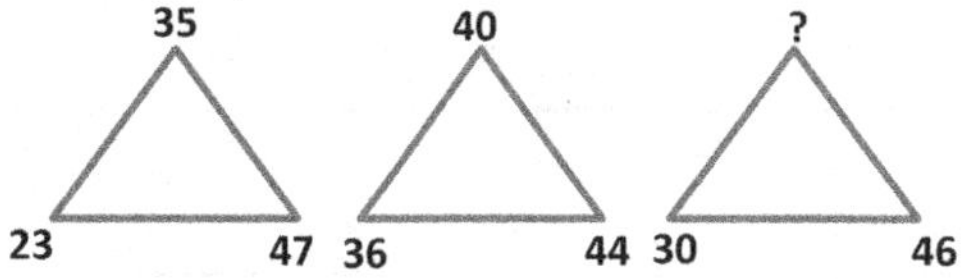

3) How many rectangles of any size are in this diagram?

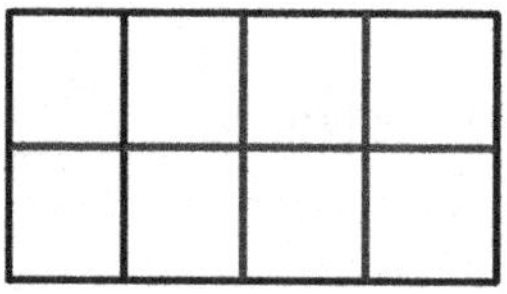

4) What is the perimeter of a rectangle with an area of 48 cm^2 and a width of 3 cm?

5) Andrew used toothpicks to build this line of connecting triangles. How many toothpicks will he need to make 80 triangles?

6) Jim is twice as old as Todd. Todd is three times as old as Bart. The sum of all three ages is 90 years. How old is Todd?

7) Jim, Bob, and Carl have money for the concession stand. Jim and Bob have $15 combined. Bob and Carl have $16 combined. Jim and Carl have $13 combined. How much money do Jim, Bob, and Carl have altogether?

8) In this magic square, five more numbers can be placed in the boxes so that the sum of the three numbers in each row, each column, and in each diagonal is the same amount. What is the value of N?

16		12
		N
8	18	

9) Arnold and Ken each threw darts at a target. Arnold hit region X four times and region Y two times for a score 42 points. Ken hit region X three times and region Y three times for a score of 45 points. How many points was region X worth?

10) Judy put her cupcakes in 16 equal rows. If she rearranges 4 rows , and moves those cupcakes so that there are 12 rows instead of 16, then each of those 12 rows would get 4 extra cupcakes. How many cupcakes does she have?

TEAM ANSWER DOCUMENT
(Grades 5-6)

SCORE

Team # __________ School ______________________________

Student Names ______________________________

____

Write answers clearly. Each correct answer is worth one point.

1. __________
2. __________
3. __________
4. __________
5. __________
6. __________
7. __________
8. __________
9. __________
10. __________

PERENNIAL MATH

Grade 5-6 Team Event Solutions

1) The sum of first seven prime numbers is 2 + 3 + 5 + 7 + 11 + 13 + 17 = 58

2) The top number in each triangle is the average of the 2 bottom numbers. The average of the third triangle is 30 + 46 divided by 2 = 38.

3) How many rectangles of any size are in this diagram?

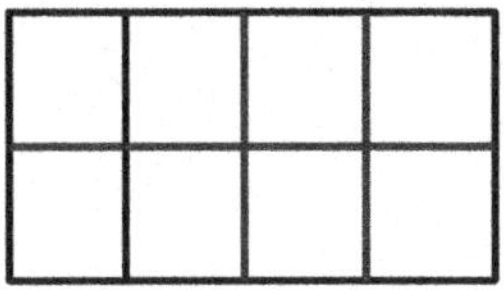

Type	How Many
Made of 1 shape	8
Made of 2 shapes	10
Made of 3 shapes	4
Made of 4 shapes	5
Made of 6 shapes	2
Made of 8 shapes	1
	30

4) The area of a rectangle is length x width. Since the width is 3, the length must be 16 cm in order for the area to be 48 cm^2. To get the perimeter add up the sides 3 + 16 + 3 + 16 = 38 cm

5) You could draw these out and count or if you make a chart you can see the relationship of triangles to toothpicks and deduce the formula that toothpicks = 2 x number of triangles +1. Therefore, for 80 triangles you need 2 x 80 + 1 = 161 toothpicks.

Number of Triangles	1	2	3	41
Number of Toothpicks	3	5	7	83

Grade 5-6 Team Event Solutions

6) Make a chart to compare their ages: Todd is 27 years old.

Jim (2T)	Todd (3B)	Bart	Total (90)
30	15	5	50
36	18	6	60
42	21	7	70
48	24	8	80
54	27	9	90

7) Use Algebra

$$\begin{array}{rcl} J + B & & = 15 \\ B + C & & = 16 \\ \underline{J + \quad C} & & \underline{= 13} \\ 2J + 2B + 2C & & = 44 \end{array}$$

Divide by 2 and J + B + C = 22. Now you can substitute the original equations and deduce that 15 + C = 22, So C = 7. Substitute C in and B + 7 = 16, therefore B is 9. Substitute 9 in for B and get J + 9 = 15. Therefore J is 6.

8) Pick a corner that is shared by two sides so that you can do algebra. The bottom right corner is shared by the bottom row and the right column. So row (8 + 18 + x) = (12 + N + x). Simplify and you get (8 + 18) = (12 + N). The missing number is 14. Corner sums are equal. So (8 + 12) = (16 + x). The bottom box x = 4. Now you can solve for a side (8 + 18 + 4) = 30. The middle box is equal to (8 + m + 12) =34. The middle box m is equal to 10.

16		12
		N
8	18	X

9) Use algebra: (Arnold) 4x + 2y = 42 (Ken) 3x + 3y = 45

3x + 3y = 45 divide both sides by 3 and you get x + y = 15; therefor 2x + 2y = 30.

Subtract this from Arnold's 4x + 2y = 42

$$\begin{array}{r} \underline{-2x + 2y = 30} \\ 2x \qquad = 12 \\ x = 6 \end{array}$$

10) You could draw this out and have 16 rows. Since you are moving 4 rows out to leave 12 rows and those 12 rows are each receiving 4 cupcakes added to them, then there were 48 cupcakes moved from those 4 rows. You can deduce that each row had to have had 12 in it to start with since you moved 48 cupcakes and they made up 4 rows. 16 original rows times 12 cupcakes equal 192 cupcakes.

PERENNIAL MATH

TEAM ANSWER KEY

(Grades 5-6)

Instructions for Scoring:

- ***Mark CORRECT answers by circling "O" the question number, and INCORRECT answers with an "X" . Each correct answer is worth one point. Count the circles (correct answers) and write that number in the "SCORE" box.***

1. **58**
2. **38**
3. **30**
4. **38cm**
5. **161**
6. **27**
7. **$22**
8. **14**
9. **6**
10. **192**